P-51 Mustang

By Larry Davis

Color By Don Greer

squadron/signal publications

Major George Preddy was the top scoring Mustang ace in the European Theater of Operations. He flew many aircraft, all named *CRIPES A' MIGHTY*, including this P-51D, his last Mustang. He was based at Bodney during November of 1944. The P-51B in the backgrouand was flown by Glennon Moran, an ace with thirteen kills.

"Dutch" Kindelberger, President of North American Aviation. (NAA)

COPYRIGHT 1995 SQUADRON/SIGNAL PUBLICATIONS, INC.
1115 CROWLEY DRIVE CARROLLTON, TEXAS 75011-5010
All rights reserved. No part of this publication may be reproduced, stored in a retrieval system or transmitted in any form by means electrical, mechanical or otherwise, without written permission of the publisher.

ISBN 0-89747-350-7

If you have any photographs of aircraft, armor, soldiers or ships of any nation, particularly wartime snapshots, why not share them with us and help make Squadron/Signal's books all the more interesting and complete in the future. Any photograph sent to us will be copied and the original returned. The donor will be fully credited for any photos used. Please send them to:

Squadron/Signal Publications, Inc.
1115 Crowley Drive.
Carrollton, TX 75011-5010

Если у вас есть фотографии самолетов, вооружения, солдат или кораблей любой страны, особенно, снимки времён войны, поделитесь с нами и помогите сделать новые книги издательства Эскадрон/Сигнал ещё интереснее. Мы переснимем ваши фотографии и вернём оригиналы. Имена приславших снимки будут сопровождать все опубликованные фотографии. Пожалуйста, присылайте фотографии по адресу:

Squadron/Signal Publications, Inc.
1115 Crowley Drive.
Carrollton, TX 75011-5010

軍用機、装甲車両、兵士、軍艦などの写真を所持しておられる方はいらっしゃいませんか？どの国のものでも結構です。作戦中に撮影されたものが特に良いのです。Squadron/Signal社の出版する刊行物において、このような写真は内容を一層充実し、興味深くすることができます。当方にお送り頂いた写真は、複写の後お返しいたします。出版物中に写真を使用した場合は、必ず提供者のお名前を明記させて頂きます。お写真は下記にご送付ください。

Squadron/Signal Publications, Inc.
1115 Crowley Drive.
Carrollton, TX 75011-5010

Acknowledgements

Joe Bruch	A.C. Chardella
J. V. Crow	Jeff Ethell
Bob Esposito	Don Garrett
John Horne	Tom Ivie
Fred Johnson	Robert Kuhnert
Jim Lansdale	Ernie McDowell
Dave McLaren	Kieth Melville
Dave Menard	MGEN Stanley Newman
Merle Olmsted	Rockwell International (NAA)
Dick Starinchak	Jim Sullivan
COL Richard Turner	USAF Museum
Paul Vercammen	Richard L. Ward
Nick Williams	Nicholas J. Waters III

Overleaf: The ultimate fighter aircraft of the Second World War, a P-51D Mustang. (NAA)

Introduction

MUSTANG - just the word brings to mind high performance. And during the Second World War, it was the nickname for what is generally acknowledged to be one of the finest fighter aircraft of the war. It was known by other "official" names like Invader and Apache, and unofficial terms like SPAM CAN. The aircraft had many shapes, different engines, and one variant even had two fuselages. But they were all still Mustangs, "The most aerodynamically perfect pursuit plane in existence!"

The design of the Mustang began as a series of RAF requirements for a fighter aircraft that could be built in the United States to augment the dwindling fighter inventories being used up in the life and death struggle of the Battle of Britain. Indeed, the Mustang design and developments are directly tied to British requirements throughout the history of the airplane. The RAF needed a new fighter and North American Aviation gave them the Allison-engined P-51/Mustang I series. When the RAF wanted to develop the design further through use of a higher performance power plant, North American offered them the Rolls-Royce Merlin powered P-51B Mustang series. The use of the sliding bubble canopy was based on the RAF developed Malcolm Hood; although the 360 degree, full vision canopy was a purely North American design.

The Mustang was designed and developed throughout the Second World War to meet various Allied air force requirements for a long range escort fighter aircraft. It more than met those requirements as many a Luftwaffe pilot will attest. But after the end of the war, when the other major fighter types like the P-38, P-47, and Spitfire, were long gone from air force inventories, the Mustang was still flying combat. Five years after the end of the Second World War, the Mustang was the leading fighter-bomber in the "jet war" in Korea. Four years after the end of the Korean War, Mustangs were fighting over the Sinai Desert with the Israeli Defense Force/Air Force. Mustangs were the basis of defense by more nations than any other major fighter type in history. Ironically, the only two major nations not to be equipped with Mustangs were the old foes of the P-51, Germany and Japan.

Today only a handful of the 16,000+ Mustangs that were built survive. Some are in museums, some are flying in air races, many fly the air show circuit and some are movie stars. But all those lucky enough to see them flying still marvel at, "The most aerodynamically perfect pursuit plane in existence."

Allison Engined Mustangs

The first Mustang grew out of a British requirement for a second manufacturer to license-build the Curtiss P-40 Tomahawk. But the people at North American convinced the British Purchasing Committee that they could design and build a better aircraft than the P-40. Slightly over four months later the NA-73X prototype was rolled into the California sun. The British named the aircraft the Mustang I.

RAF MUSTANG Is began coming off the Inglewood assembly line during mid-April of 1941. Powered by the tried and tested Allison 1,120 hp, twelve cylinder, supercharged V-1710 engine, the Mustang I could attain a top speed of 382 mph at 13,000 feet. Tested and found more than suitable by U.S. Army Air Force pilots, the U.S. Army purchased 150 of the new aircraft, designating it P-51. The difference between USAAF P-51s and RAF Mustang Is was mainly in the armament. The Mustang I had a pair of .50 cal. guns in the nose, and two more .50s and four .303 guns in the wings. The P-51 was armed with four 20MM can-

The XP-51 prototype on the ramp at Mines Field, following rollout ceremonies in the Summer of 1941. The XP-51s were actually RAF-specified Mustang Is that were used for test and evaluation by the U.S. Army Air Corps. (NAA)

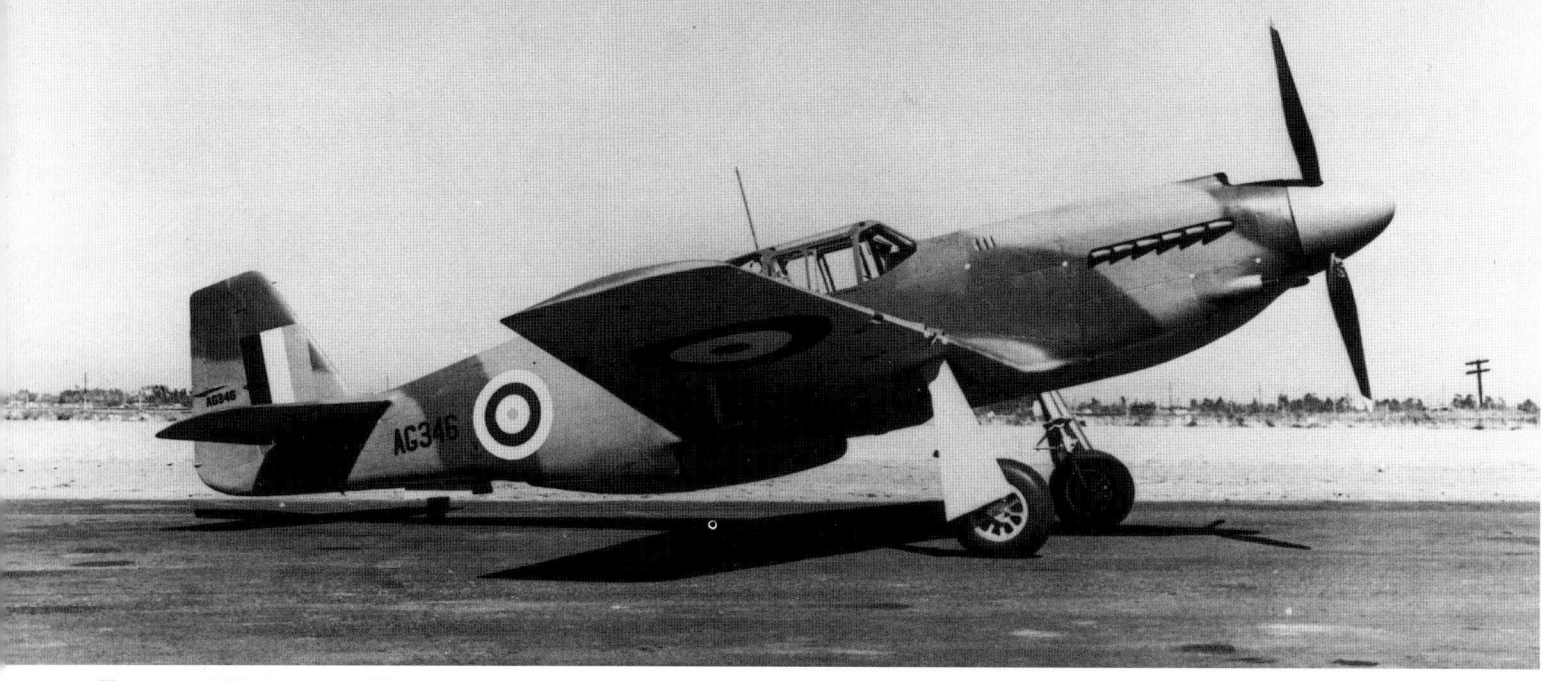

The second Mustang I at Mines Field during May of 1941. This aircraft, AG 346, was painted in RAF camouflage, tested, disassembled, crated up and sent to England in September of 1941, for RAF flight testing. (NAA)

non. Later, the RAF bought additional Mustangs identical to the 20MM armed P-51, designating the type Mustang IA.

RAF Mustang Is were the first to see combat, with No 2 Squadron becoming operational during April of 1942. In July of 1942, the first RAF fighter aircraft penetrations into Germany began. The only aircraft that could do it was the Mustang I. Eventually twenty-eight RAF/RCAF/RNZAF/SAAF squadrons were equipped with Mustang I/IA aircraft for either the fighter or reconnaissance mission. The British

The Army Air Corps production version of the Mustang I was designated P-51, and differed from the Mustang I in having an armament of four 20MM cannon in the wings. (NAA)

purchased over 700 Mustang I and IA aircraft during the war.

The U.S. Army Air Force bought 150 of the P-51s, nicknamed Apache, most of which had reconnaissance cameras behind the cockpit and below the rear fuselage. These aircraft were 'officially' designated F-6A, although the data block still read P-51. The first U.S. unit to take the P-51 into combat was the 154th Observation Squadron in Tunisa. They flew their first mission during April of 1943 against targets in Sicily. Only two AAF squadrons flew the P-51/F-6A in combat, the 154th OS and the 111th TRS.

The A-36A, nicknamed Invader, was a dive bomber variant of the P-51. It differed from the P-51 in having a set of dive brake panels on the upper and lower wings, and in having the wings strengthened to be able to carry a pair of 500 pound bombs. Additionally, the wing armament was changed to four .50s, with another pair of .50s in the nose.

Although the AAF purchased 500 A-36As, only three combat units

A staged posing of a new production RAF Mustang I at Mines Field (now the site of LA International) during 1942. The Mustang I was delivered in the standard RAF camouflage of Dark Earth and Dark Green, with Sky undersurfaces. The lengthened carburetor scoop was an identification feature of all production Allison-engined Mustangs. (NAA)

were equipped with the type. Most of the production run ended as dive-bombing trainer aircraft. The 27th FBG at Rasel Ma, French Morocco was the first unit to fly the A-36A in combat, with their first mission coming during June of 1943. Two other AAF groups flew the A-36A in combat, the 86th FBG in the Mediterranean Theater, and the 311th FBG in the China/Burma/India Theater. The RAF had no A-36A squadrons,

Mustang Is entered RAF service in April of 1942. This Mustang I (AM 148) was assigned to No 26 Squadron. The aircraft was enroute to the disastrous Dieppe landings in France in August of 1942. (via R.L. Ward)

All the Mustang Is were built at North American's Inglewood plant, test flown, then disassembled, crated up, and sent to England by ship. This Mustang I was being reassembled at RAF Burtonwood. (NAA)

This Mustang I (AM 106) was used by the RAF at Boscombe Down for the flight testing of underwing drop tanks, bombs and large caliber cannons used for tank busting. (Imperial War Museum via R.L. Ward)

Ground crews push *Betty Jean*, a P-51 assigned to the 111th Tactical Reconnaissance Squadron, into a sand bag and oil drum revetment at Anzio in April of 1944. Army Air Corps P-51s were operational in North Africa beginning in April of 1943. (USAAF)

A mechanic adjusts the camera mounted behind the cockpit of a 111th TRS P-51 during 1944. By this time, the P-51 had been re-designated as the F-6 by the Army Air Corps. (James V. Crow)

although the type was tested.

The final variant of the Allison Mustang was the P-51A/Mustang II. The P-51A incorporated all the changes developed through previous types. Both AAF and RAF aircraft were similarly equipped, with the reinforced A-36A wings minus dive brake panels, and armed with four .50s in the wings. The nose guns were deleted. Horsepower and propeller changes took the top speed to 409 mph at 10,000 feet. North American built 310 P-51As, with 50 being purchased by the RAF and designated Mustang II. Many of the P-51As were also equipped with cameras for the recon mission. These aircraft were designated F-6B in the AAF.

The first P-51As in combat were with the 311th FBG based at Dinjan, India, during mid-1943. The 311th FBG had two squadrons equipped with P-51As, and the third equipped with A-36As. Two other units in the CBI Theater were equipped with the P-51A, the 23rd Fighter Group "Flying Tigers", and the 1st Air Commando Group. Both the 23rd FG and 1st ACG flew long range escort missions deep into China and Formosa with the P-51As. Several camera-equipped F-6Bs flew with 9th AF units in Europe during the Summer of 1944. The RAF had two squadrons of Mustang IIs, although both were primarily equipped with Mustang I/IA aircraft and the Mustang IIs simply filled combat losses.

Captain Paul Hexter (center) designed the disruptive Black and White "dazzle" camouflage pattern being sprayed on this P-51 Mustang. The pattern was applied to the fuselage sides and the undersides of the wings and tail surfaces only. (USAAF)

A Mustang I of No 2 Squadron, RAF. The RAF had twenty-eight squadrons equipped with Mustang I & IA aircraft. RAF Mustangs were used mainly in the fighter-bomber and photo reconnaissance roles. (Ernie McDowell)

(Right) This ex-RAF Mustang I (AG348) was the fourth production Mustang I. The aircraft was supplied to the Soviet Union for testing during 1942. (via Hans-Heiri Stapfer)

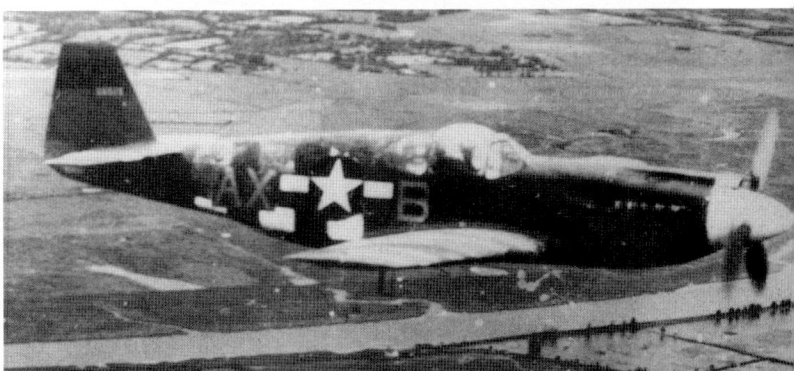

Lieutenant Colonel Russ Berg at the controls of *Little Sir Echo*, an F-6 assigned to the 107th TRS, over Europe in July 1944. The aircraft has overpainted D-Day bands on the wing upper surfaces and has a Malcolm bubble canopy. (Colonel R.S. Stone)

A pair of P-51s fly side-by-side to show off the underside pattern of Captain Paul Hexter's disruptive paint scheme. The upper surfaces were painted standard Olive Drab. The "dazzle" scheme was meant to disrupt an enemy gunners aiming point, but was never adopted for operational use. (USAAF)

Lieutenant Colonel James Deering stands beside *Richfield Rowdy*, the P-51 he flew when he commanded the 154th Observation Squadron in North Africa during 1943. (James V. Crow)

An P-51 from the 111th Tactical Reconnaissance Squadron, "Snoopers", on the ramp of its home base in Italy during 1944. The aircraft has Yellow bands around the wings and tail surfaces, which were used to help identify the P-51 from a Bf 109 in combat situations. (USAAF)

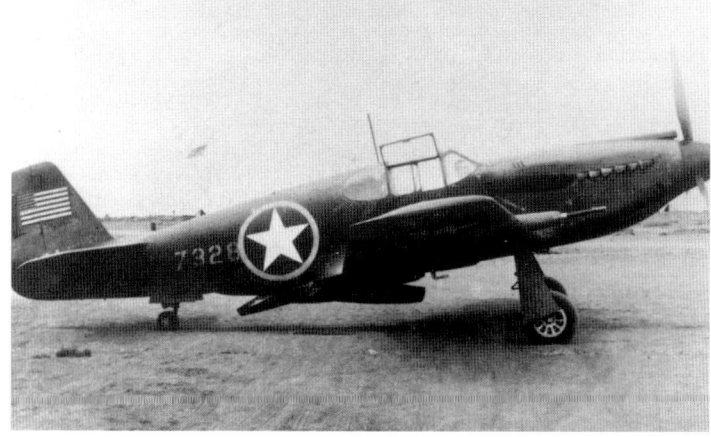
The first Army Air Force unit to take the P-51 into combat was the 154th Observation Squadron during April 1943. This P-51 still carries the U.S. Flag and Yellow surround to the national insignia carried on all aircraft involved in Operation TORCH, the invasion of North Africa. (Warren Bodie)

A RAF camera-equipped FR-1A Mustang during 1943. The aircraft was camouflaged in the later Dark Green and Dark Gray used for northern European operations. RAF Mustang Is flew the first fighter penetration of Germany on 27 July 1942. (via Jeff Ethell)

An A-36A Invader of the 27th Fighter Bomber Group banks toward Mount Vesuvius during 1943. Since it carries no unit markings, this A-36A was probably a new replacement aircraft in the 27th FBG. (USAAF)

The A-36A Invader was a P-51 with upper and lower wing dive brakes and reinforced underwing pylons to carry 500 pound bombs. The A-36A was developed for use as a dive bomber and close support aircraft. (NAA)

A-36A Invader pilots received their dive-bombing training at Baton Rouge Army Air Base. This pilot trainee prepares for a practice mission carrying 100 pound practice bombs. (via Jeff Ethell)

An A-36A Invader from the 27th Fighter Bomber Group on the ramp at Rasel Ma in French Morocco during the Spring of 1943. The 27th Fighter Bomber Group was the first combat operational unit in the Apache. (USAFM)

(Left & right) *Doodle/El Matador* was an 86th Fighter Bomber Group A-36A flown by F/O R.L. Bryan (seated on wing). F/O Bryan flew over forty missions against Axis targets from bases in Sicily during 1943. (USAFM)

This 27th FBG A-36A sits on the grassy ramp at Gela, Sicily during late 1943. The Invader has 150 mission markings, ninety-eight dive bombing, forty-two ground strafing, and ten armed reconnaissance missions. The camouflage is extremely weathered and a Blue surround was added to the national insignia. (USAAF)

The RAF bought a single A-36A Invader but rejected the type in favor of the Typhoon/Tempest series. Combat tests of the RAF A-36A were flown alongside Mustang I squadrons at Foggia, Italy. The dive brake under the left wing was deployed. (via Jeff Ethell)

Maintenance conditions in the Mediterranean Theater were crude even when they were at their best and rain often turned an airfield into a sea of mud. This 27th FBG A-36A was being set up for bore sighting of the guns. The aircraft carries two Yellow swastika victory makings painted on the nose. (USAAF)

This 86th FBG A-36A, based at Tafaraoul Air Base, Algeria, carries the name *Hellsa' Droppin* inside the bomb outline painted on the nose. This A-36A has had the twin .50 caliber machine guns removed from the nose. (via Jeff Ethell)

At least one additional A-36A found its way into the RAF. HK 944 was assigned to the Communications Squadron at Foggia during November of 1943. The "patches" on the fuselage are gas detector panels. (Howard Levy via R.L. Ward)

The pilot of this veteran 27th Fighter Bomber Group A-36A with sixty-three mission markings, took a hit over Sicily but made it back to its base in Morocco and made a belly landing. (USAAF)

Armorers load 500 pound bombs under the wings of an 86th Fighter Bomber Group A-36A during 1944. A-36As, also known as Invaders, flew bridge-busting missions in Italy as part of Operation STRANGLE. (USAAF)

Dive bombing pilot trainees learned their trade at Baton Rouge Army Air Base, in first the Douglas A-24 Dauntless, then in the A-36A. They then took the A-36A into combat. (Dick Starinchak)

A-36As from the 27th Fighter Bomber Group line the ramp at Gela, Sicily during 1944. Each aircraft carries a minimum of 100 mission marks. The Invader was the primary fighter-bomber in the MTO during 1943. (NAA)

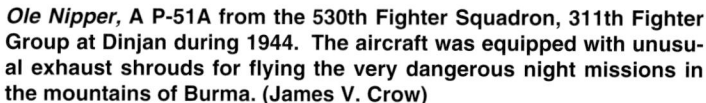

Mechanics tune the engine on Captain John England's *Jackie,* a P-51A from the Yellow-nosed 530th Fighter Squadron based at Dinjan, India. The P-51A was a fighter version of the A-36A minus the dive bombing brakes and bombing equipment. (USAAF)

Ole Nipper, A P-51A from the 530th Fighter Squadron, 311th Fighter Group at Dinjan during 1944. The aircraft was equipped with unusual exhaust shrouds for flying the very dangerous night missions in the mountains of Burma. (James V. Crow)

Ferry pilots prepare a P-51A for delivery during 1943. P-51As equipped only three Army Air Force units, the 311th and 23rd Fighter Groups, and the 1st Air Commandos, all in the China-Burma-India Theater. (via Nick Williams)

The five White stripes around the fuselage of this pair of P-51As, indicate that they were assigned to the 1st Air Commando Group. The 1st ACG provided fighter escort for 10th Air Force B-25s attacking Japanese targets throughout Burma and China. (USAAF)

A P-51A from the 1st ACG on the ramp at Hailakandi, India, during March of 1944. P-51As flew the first very long range bomber escort missions when they escorted 10th Air Force B-25s to targets on Formosa. (R.C. Jones)

1st Air Commando P-51As line the ramp at Broadway Strip, Burma during 1944. The Air Commandos were a composite group having complete fighter, bomber and cargo squadrons attached. The propeller spinner and fin tip were in Natural Metal. (USAAF)

Maintenance in the CBI Theater went through some of the worst conditions in the history of air warfare. Always working in the outdoors, these 1st ACG crew chiefs begin refueling *"Mrs. Virginia"* at Hailakandi, India. (USAAF)

A 311th Fighter Group crew chief makes some adjustments to the big Allison V-1710 at Kurmatola, India, following some battle damage incurred while escorting B-25s to Formosa during late 1943. (USAAF)

A 1st ACG P-51A sits on the Dinjan ramp armed with a pair of 500 pound bombs and six underwing 5 inch Bazooka rocket launchers during August of 1944. (USAAF)

Jeanie, an F-6A from the 107th TRS on the ramp at Le Molay Airdome, France, during the Summer of 1944. The aircraft had thirty-two mission markings on the nose in White and was equipped with a Malcolm bubble canopy, which could be and was fitted to many early Mustang variants. (Fred LePage)

Merlin Engined P-51B

The P-51B resulted from a joint British/U.S. Army Air Force program to increase the overall performance of the Mustang. North American could have opted for further development of the Allison V-1710. But time and the war couldn't wait for such a long term project. The only other engine readily available, that could produce the kind of horsepower needed to make the Mustang a truly superb fighter aircraft, capable of beating the Luftwaffe anywhere over Europe, was the British Rolls Royce V-1650 Merlin, already in production for use by Royal Air Force Spitfires, Hurricanes, and Lancaster bombers.

The big Merlin, with its two-speed, two-stage supercharger, put out almost 1,400 horsepower. Installation of the Merlin, license-built in the U.S. by Packard Motor Car Co., took the top speed of the Mustang to 430 mph at 25,000 feet. It was that altitude factor that made the big difference. The Allison Mustang developed its greatest horsepower 15,000 feet below that of the Merlin powered variant. The Merlin-powered Mustang could now compete with the best the Luftwaffe had, at the altitudes the Germans flew at. The first Merlin-powered Mustang flew during October 1942.

After almost a full year of development and testing, redesign and retesting, the Merlin Mustang, designated P-51B in the U.S. Army Air Force and Mustang III in the Royal Air Force, was ready for combat. The first units to receive the P-51B were quite naturally in the ETO. During early December of 1943 the 354th Fighter Group, "The Pioneer Mustang Group", flew its first combat mission. Although this first mission was a short familiarization sweep into France, it wasn't long before the 354th FG Mustangs were staying with the bombers well into Germany, far beyond the range of other Allied fighter types. On 4 March 1944, P-51B Mustangs from the 4th Fighter Group escorted the bomber force all the way to Berlin and back. Hermann Goering would remark after the war that it was on that day that he knew Germany could not win the war.

Royal Air Force squadrons began receiving Mustang IIIs soon after U.S. Army Air Force units, many of them replacing veteran Spitfires and Hurricanes. Neither British type had the range necessary for the long range escort missions into Germany. North American built 308 Mustang IIIs for the RAF units. The first RAF unit in operation with the Mustang III was No 19 Squadron, which flew its first missions during February of 1944.

So great was the demand for Merlin Mustangs that North American opened a second assembly line at its new plant in Dallas, Texas. These aircraft were designated P-51C. Many P-51B/C aircraft had cameras installed behind the cockpit and under the rear fuselage. These carried the Army Air Force designation of F-6C. The RAF received both P-51B and P-51C aircraft, but the designation remained Mustang III for both types. Mustang IIIs eventually equipped thirty squadrons in the RAF. Army Air Force units in all theaters save the Pacific Theater, converted to P-51B/Cs as soon as production aircraft were available. North American produced a total of 3738 P-51B/C aircraft.

North American engineers modified several P-51 airframes to accept the Packard-built Rolls-Royce Merlin engine. This P-51 retained its four 20MM cannon, while production P-51Bs were armed with four .50 caliber machine guns. The deep bulge under the nose was found only on prototype Merlin Mustangs. (NAA)

The first operational group equipped with P-51Bs was the 354th Fighter Group, based at Boxted during November of 1943. This 356th FS P-51B carries typical ETO Mustang identification markings designed to allow a quick combat identification. These consisted of a White nose and spinner, with White bands around the wings and tail surfaces. (USAAF)

Typhoon McGoon was a P-51B of the 363rd Fighter Squadron, 357th Fighter Group, based at Leiston, England. The 357th Fighter Group flew their first P-51B mission during February of 1944. (Ernie McDowell)

Although the Mustang had plenty of range to make the ferry flight from the U.S. to bases in Great Britain, the majority of all P-51 types were shipped by sea. These P-51Bs were being off-loaded from a U.S. Navy CVE in Liverpool, England during February of 1944. (via Jeff Ethell)

Lieutenant William Hovde flew *Ole-II*, a P-51B of the 358th Fighter Squadron at Steeple Morden. The scoreboard shows forty-eight P-47 missions before conversion to the P-51B in March of 1944. Hovde's P-51B has a pair of Spitfire rear view mirrors mounted on top the windscreen. (Keith Melville)

Meiner Kleiner, a P-51B from the 334th Fighter Squadron, 4th Fighter Group at Debden, was flown by Lieutenant Joe Higgens. The 4th Fighter Group converted to P-51Bs in late February of 1944, immediately following the Big Week operation. (Fred Dickey Jr.)

Hot Pants was the P-51B flown by Lieutenant Richard Rabb when he was assigned to the 370th Fighter Squadron. The aircraft carried twelve victory markings under the windscreen. *Hot Pants* was parked on a Swedish air base where another pilot landed after suffering some combat damage over Germany in the Summer of 1944. (Ernie McDowell)

Lieutenants Franklyn Hendrickson, William Pitcher, and Edward Phillips stand by a P-51B of the 362nd Fighter Squadron following an escort mission to Brunswick, Germany in early 1944. (Keith Melville)

The *Jovial Judge* was a Mustang III assigned to No 260 Squadron, Royal Australian Air Force. The aircraft was flown by Sergeant Tom Ferguson during September of 1944. No 260 Squadron was part of the RAF Desert Air Force. (John Horne)

Lieutenant Henry Brown's second *The Hun Hunter From Texas* (WR*Z), has thirteen of his victories painted on the fuselage under the cockpit. He gained his kills while flying with the 354th Fighter Squadron. (Bill Marshall)

Ding Hao! was flown by Medal Of Honor recipient Major James Howard. The scoreboard on his 356th Fighter Squadron P-51B shows the eight and a half Japanese victories scored while flying with the Flying Tigers, and six German victories gained while with the 354th Fighter Group. (Colonel Richard Turner)

City Of Paris/Hoo Flung Dung was a P-51B flown by Major Robert McWherter, of the 363rd Fighter Squadron. Major McWherter's P-51B was equipped with one of the sliding, Malcolm Hood bubble canopys developed by the RAF for the P-47 and P-51. (Ernie McDowell)

This very strange P-51B was loaned to the N.A.C.A. for some, as yet, unexplained flight test. It carried Orange and Black identification bands. (Dave McLaren)

Several P-51Bs were shot down inside Germany and landed with repairable damage. A few were repaired and returned to flight status with the Luftwaffe, being used as trainers for German pilots to test their skills, and to develop tactics to be used against the P-51. (Ernie McDowell)

Texas Terror III, was a 354th Fighter Squadron P-51B flown by Lieutenant Lee Mendenhall. It carried four German victory markings and eight bombing mission markings during the Spring of 1944. (355th Fighter Group Assn.)

W/O J. Quinn flew this Mustang III when he was assigned to No 3 Squadron, Royal Australian Air Force. Most of the missions flown by Desert Air Force units were fighter-bomber sorties. For this role, the normal ordnance load was two 500 or 1,000 pound bombs. (John Horne)

Tommy's Dad was John "Pappy" Herbst. Pappy Herbst scored eighteen victories when he flew this P-51B with the 74th Fighter Squadron based at Liuchow, China during 1944. (Tom Ivie)

The 23rd Fighter Group, the "Flying Tigers" turned in their venerable P-40 Warhawks for P-51B Mustangs during 1943. This sharkmouthed P-51B had Yellow and Black tail bands, with a White nose and wing tips. The unit was based at Liuchow, China during November of 1944. (USAAF)

Colonel David "Tex" Hill climbs into the P-51C he flew when he commanded the 23rd Fighter Group during October of 1944. Colonel Hill had twelve and a half victories with the American Volunteer Group during 1942. The Mustang's camouflage was extremely worn and weathered, which was typical of all CBI aircraft. (USAAF)

Lieutenant John Stricker taxis The *Green Hornet*, a P-51B of the 382nd Fighter Squadron, from its parking spot at Staplehurst during the Spring 1944. His crew chief, SGT John Ross, rides the wing to guide Strcker to the runway. There are two extra "gun" ports painted on the wings. (USAAF)

Captain Don Gentile (left) beside his famous P-51B, *Shangri-La* during the Spring of 1944. Gentile would score over thirty air and ground victories while flying with the 336th Fighter Squadron. (Keith Melville)

Showing forty-four missions on the nose, this 358th Fighter Squadron P-51B crashed at 8th AF Station F-375, Honnington, on 18 July 1944. The Black and White D-Day bands have been crudely over-painted on the upper surfaces only. (USAAF)

Lieutenant Elmer Cater flew this P-51B with the 368th Fighter Squadron during May of 1944. Cater took several accurate hits in the radio compartment shattering the rear canopy glass and several others in the aft fuselage, but still managed to bring the aircraft home. The aircraft carried a single kill marking under the windscreen. (A.C. Chardella)

Lieutenant Donald Bochkay flew *Speedball Alice* with the 363rd Fighter Squadron, 8th Air Force at Yoxford. Bochkay scored thirteen and a half victories. (James V. Crow)

Major Claibourne Kinnard (middle) flew *Man O' War/The Bulldog 1*. He was an ace with at least eight victories gained while he was assigned to the 354th Fighter Squadron, 8th Air Force Fighter Command, based at Steeple Morden. (Bill Marshall)

These sharkmouthed Mustang IIIs were assigned to No 112 Squadron, RAF Desert Air Force while based at Lavariano, Italy during 1945. The Royal Air Force had thirty squadrons equipped with the Mustang III. No 112 Squadron aircraft had a long tradition of carrying sharkmouth markings, beginning with their Tomahawks (P-40Bs). (R.A. Brown via R.L. Ward)

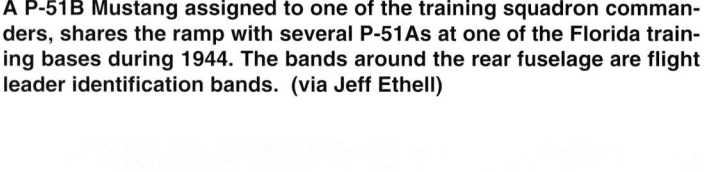

A P-51B Mustang assigned to one of the training squadron commanders, shares the ramp with several P-51As at one of the Florida training bases during 1944. The bands around the rear fuselage are flight leader identification bands. (via Jeff Ethell)

A P-51B from the 352nd Fighter Group, the "Blue-Nosed Bastards From Bodney". The underlined 'R' on the fuselage codes indicates the Mustang was the second aircraft in the squadron to carry the individual identification letter 'R'. (USAFM)

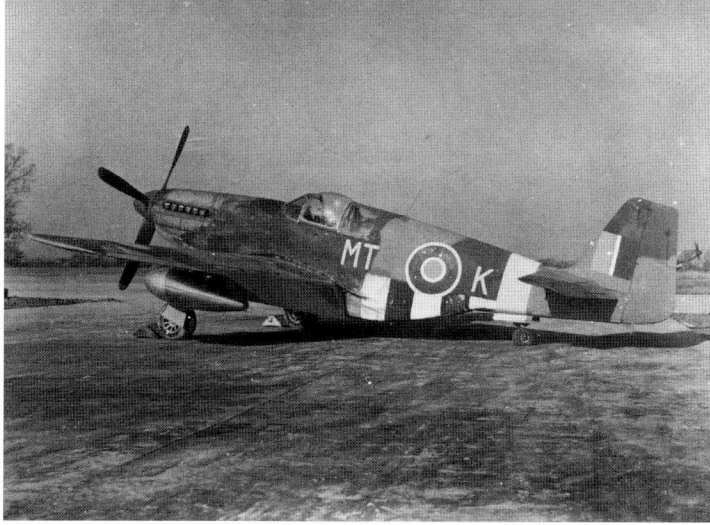
An RAF Mustang III of No 122 Squadron. The upper surface D-Day bands have been re-camouflaged by July of 1944. (Ernie McDowell)

Bullfrog was a P-51C assigned to the 51st Fighter Group at Kunming, China during 1944. The P-51C was a P-51B model built at the North American plant in Dallas, Texas. (Ernie McDowell)

(Right) Carrying full D-Day invasion markings, PZ*S of the 486th Fighter Squadron was based at Bodney in June of 1944. The D-Day stripes were usually quite crudely brush painted on, not masked and sprayed. (Ernie McDowell)

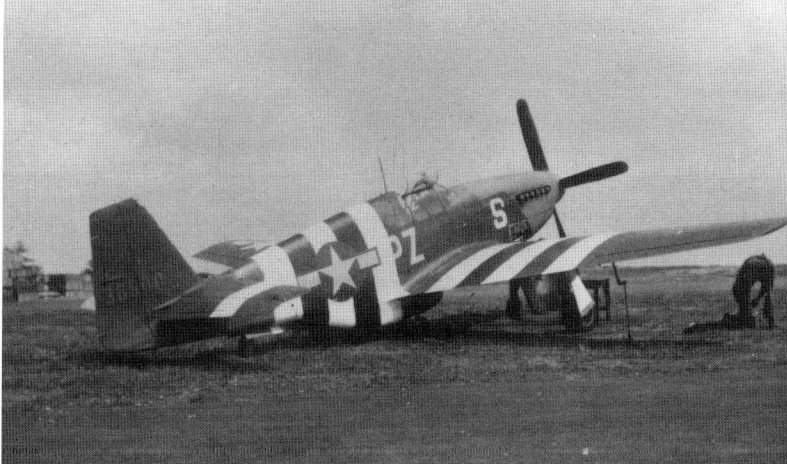

This pair of F-6C Mustangs are assigned to the 10th Tactical Reconnaissance Group, 9th Air Force, based at Lessay, France during 1944. (Joe Scogna via Jeff Ethell)

Easy Rockin' Mama, was a P-51B of the 353rd Fighter Squadron. The aircraft was covered by camouflage netting at Lashenden in the late Spring 1944, just prior to D-Day. (USAAF)

U've Had It! was certainly correctly named since its 362nd Fighter Squadron pilot, Captain John England, had nine kill markings under the cockpit during the Summer of 1944. (Merle Olmsted)

Betty Jane was one of Colonel Charles McCorkle's Mustangs. He flew this aircraft when he commanded the 31st Fighter Group and carried ten kill marking on the fuselage side. Based at San Severo, Italy, the 31st Fighter Group escorted 15th Air Force B-17s and B-24s in the Summer of 1944. (Ernie McDowell)

P-51Cs from the 376th FS taxi to the active runway at Little Walden during the hectic days that followed the D-Day operations. Most of the missions flown during those first days following the invasion were in support of the ground troops, but these 361st Fighter Group Mustangs were going back to their primary mission, long range bomber escort. (USAAF)

Phyllis, an F-6C Mustang from the 12th Tactical Reconnaissance Squadron, takes off from Follonica, one of the forward air strips constructed in Italy during the Summer of 1944. (USAAF)

June Bride shares the grass ramp with other 1st Air Commando Group P-51As at Broadway Strip, Burma during 1944. (R.T. Smith via Jim Lansdale)

A well-worn 23rd Fighter Group P-51B on the ramp at Kweilin, China during 1944. The 23rd Fighter Group was descended from the old Flying Tigers, the American Volunteer Group (AVG). (USAAF)

The crew chief finishes the lettering on *The Third Dallas Blonde*, a 325th Fighter Group P-51B at Lesina, Italy during 1944. (USAAF)

(Below) *The Iowa Beaut*, was a P-51B flown by Lieutenant Lee Mendenhall of the 354th Fighter Squadron during the Summer of 1944. (USAAF)

Lieutenant Fred Allison flew *Opal Lee,* a P-51C assigned to the 487th Fighter Squadron at Bodney during 1944. It was the second aircraft in the squadron with the letter 'A', as denoted by a bar under the letter. (Fred Allison)

Major Wallace Hopkins' first *Ferocious Frankie* carried only one victory marking in June of 1944. Flying with the 374th Fighter Squadron, Major Hopkins ended the war with four victories. (Ernie McDowell)

The *Dakota Kid* was a P-51C flown by Lieutenant Nobel Peterson when he was assigned to the 358th Fighter Squadron. (355th Fighter Group Assn.)

Lieutenant Robert Powell was *The West "by Gawd" Virginian*. He was assigned to the 328th Fighter Squadron at Bodney during May of 1944. (Sheldon Berlow via Tom Ivie)

As newer P-51Ds became available, the older P-51Bs and Cs were transferred to other squadrons in the Allied air forces. This ex-23rd Fighter Group P-51B was assigned to the 11th Fighter Group, Chinese Air Force at Hsien, China during late 1944. (George McKay)

Satan's Son was an F-6C assigned to the 107th TRS at Middle Wallop during 1944. The F-6C had a camera in the left rear cockpit window, and another under the fuselage in front of the tail wheel. (Paul Miller via Merle Olmsted)

Colonel Chet Sluder flew the *Shimmy III* when he commanded the 325th Fighter Group, The Checkertail Clan, during the Russia shuttle missions of June 1944. (Ernie McDowell)

The *Texas Fire Fly*, a P-51C with the 362nd Fighter Squadron was a fine example of how the 357th Fighter Group over-painted the D-Day invasion stripes during the late Summer of 1944. (Merle Olmsted)

This RAF Mustang III was armed with underwing rocket launchers, each of which was carrying two 60 pound rockets. Since it carries no code letters, this aircraft was probably from a test and evaluation squadron. (via Jeff Ethell)

Stateside units were often very "Plain Jane" in their markings. This P-51B was assigned to a squadron commander in the 54th Fighter Group based at Bartow, Florida. It was an exception, having two Yellow squadron commander bands and a Yellow rudder. (Bob Esposito)

This F-6A of the 154th Observation Squadron. 68 Observation Group was based in North Africa during the Spring of 1943.

An A-36 Invader of the 524th Fighter Bomber Squadron, 27th Fighter Bomber Group, based at Naples, Italy during 1943.

A-36 Invaders were also active in the China-Burma-India theater. This Invader was assigned to the 311th Fighter Bomber Group at Pinjan, India during 1944.

A Mustang Ia of No 225 Squadron, Desert Air Force during 1943.

This P-51A was flown by the 1st Air Commando Group out of Broadway Strip, Burma during 1943.

Lieutenant Robert D. Brown flew this P-51B of the 362nd Fighter Squadron, 357th Fighter Group.

Captain Donald Beerbower, an ace with fifteen kills, flew this Mustang during June of 1944, while with the 353rd Fighter Squadron.

HOT PANTS carried twelve kill markings under the windscreen. The P-51B was flown by W. K. Baker of the 370th Fighter Squadron, 359th Fighter Group.

Lieutenant Colonel John Meyer, an ace with fourteen kills, flew this Blue Nosed P-51B of the 487th Fighter Squadron during May of 1944.

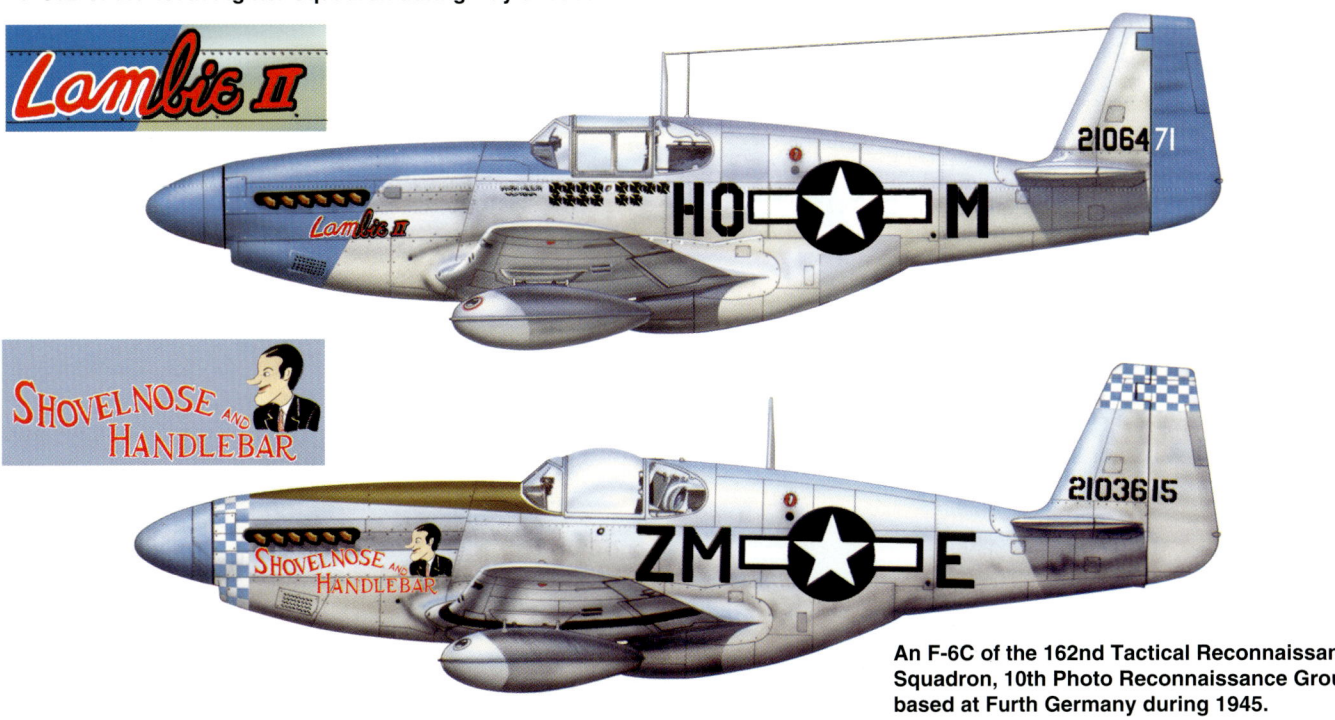

An F-6C of the 162nd Tactical Reconnaissance Squadron, 10th Photo Reconnaissance Group based at Furth Germany during 1945.

A P-51C of the 23rd Fighter Group on the ramp at Liuchow, China during late 1944. Liuchow was one of the better maintained facilities in the CBI Theater, since it had crude but effective hangers to work in. (USAAF)

This Mustang III was assigned to No 316 (Polish) Squadron, which was attached to the RAF in the Summer of 1944. The pilot of this Mustang has victory markings for eight V-1 Flying Bomb kills under the cockpit. (via R.L. Ward)

Lilli Mae, a P-51B from the 352nd Fighter Group, flies over France during September of 1944. The Mustang is unusual in that it has the individual aircraft letter on the tail rather than on the fuselage. (via Gene Stafford)

(Below) A trio of 376th Fighter Squadron P-51Cs taxi to the active runway at St. Dizier during the Summer of 1944. The second aircraft, *Elizabeth*, has a sharkmouth decoration, very unusual for a Mustang in the ETO. (Merle Olmsted)

A P-51C from the Yellow-tailed 5th Fighter Squadron based at Madna, Italy. Aircraft from the 52nd Fighter Group were unusual in that they had the same squadron codes as the 4th Fighter Group in England. (Stan Staples)

This F-6C was assigned to the 15th Tactical Reconnaissance Squadron based at St. Dizier during 1945. The aircraft had Blue and White checks on the nose and fin. (Ernie McDowell)

S/SGT Harold Staugler adjusts the landing gear oleo assembly on *Rigor Mortis*, a P-51C with the 353rd Fighter Squadron during September of 1944. (via Jeff Ethell)

(Below) *Shillelagh*, A P-51B from the 353rd Fighter Squadron, returns to Cricqueville, France, during July of 1944, after taking a hit in the oil cooler, which resulted in oil being sprayed on the rear fuselage. (USAAF)

Ferocious Frankie was flown by Lieutenant Colonel Wallace Hopkins of the 374th Fighter Squadron. The aircraft has had the D-Day markings overpainted with Olive Drab on the upper surfaces. (USAAF)

Triple Threat was a P-51D assigned to the 3rd Air Commando Squadron at Chitose, Japan, during late 1945. (Paul Vercammen)

The *Tika IV*, a 374th Fighter Squadron P-51D, was flown by Lieutenant Vern Richards during 1944. (USAAF)

(Below) *Lou IV* leads three other 375th Fighter Squadron Mustangs for a public relations flight during June of 1944. The upper surfaces are clearly Olive Drab, although many sources have mistakenly noted them as Medium Blue for many years. (USAAF)

A line-up of P-51Ds from the 23rd Fighter Group on the grass ramp at Hsein, China during early 1945. The P-47Ds in the background were assigned to the 81st Fighter Group. (George McKay)

A 3rd Air Commando P-51D on the Chitose ramp during 1945. The aircraft still carries the Black and White 5th Air Force identification bands on the wings and fuselage. (Paul Vercammen)

Marie was flown by Captain Freddie Ohr, an ace with six kills. He was assigned to the 2nd Fighter Squadron. (Fred Bamberger)

(Below) A P-51D assigned to the 11th Fighter Group, Chinese Air Force, on the ramp at Hsein, China during early 1945. (George McKay)

Lieutenant Don Lopez in the cockpit of *Lope's Hope 3rd*, the P-51C that Lopez flew when he was assigned to the 75th Fighter Squadron at Luliang, China during 1944. (Don Lopez)

Tommy, a P-51C that was assigned to the Chinese-American Composite Wing, carries dual cluster bomb units under the wings. The CACW was based at Peishyi, China during 1944. (James V. Crow)

Lieutenant Childers flew *Oh Johnie* when he was flying with the 109th Tactical Reconnaissance Squadron at Middle Wallop. The F-6C was equipped with a Malcolm bubble canopy. (James V. Crow)

Shovelnose and Handlebar was an F-6C of the 162 Tactical Reconnaissance Squadron. The aircraft was named in honor of Bob Hope and Jerry Colona, and the many USO shows they performed for the troops all over the world. The F-6 Mustangs were the airborne eyes of the 9th Air Force. (James V. Crow)

The top ace of the 325th Fighter Group was Major Herschel Green, who had a total of seventeen victories flying with the 317th Fighter Squadron. (Ernie McDowell)

Two mechanics in the 364th Fighter Squadron work on the big Merlin in this P-51B at Leiston, as a third refuels the Mustang for another mission on D-Day, 6 June 1944. (Keith melville)

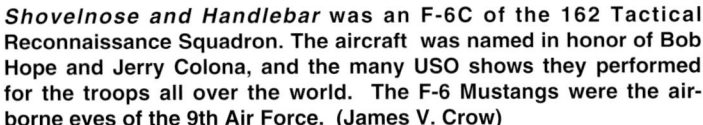

My Beth I/Sugah, an F-6C from the 109th Tactical Reconnaissance Squadron on the ramp at Strip A-87, near Gosselies, Belgium during December of 1944. The unusual fuselage code 'AI' is unexplained. (via Dave McLaren)

Five pilots from the all-Black 100th Fighter Squadron, talk about the upcoming mission in front of *Skipper's Darling III*, the P-51C flown by Captain Andrew Turner, Commanding Officer of the 100th Fighter Squadron. (USAAF)

Lieutenant Donald McKibben flew *Miss Lace*, a 486th Fighter Squadron P-51B that carried the Milton Caniff cartoon lady on the nose. (Don McKibben via Tom Ivie)

Pilot Officer Denis Evans was the pilot of this Mustang III from No 250 Sudan Squadron. The Mustang was armed with underwing launchers for 60 pound RAF-designed rockets. (Danny Morris)

HMS Hellion was flown by Lieutenant Don Whinnem when he was assigned to the 486th Fighter Squadron at Bodney. The aircraft was equipped with two 110 gallon pressed paper underwing drop tanks. (Ernie McDowell)

A Royal Air Force Mustang III fighter bomber from No 250 Sudan Squadron, enroute to targets in Italy during 1944. The fighter-bomber Mustangs could carry a pair of 1,000 pound bombs under the wings. (via Jeff Ethell)

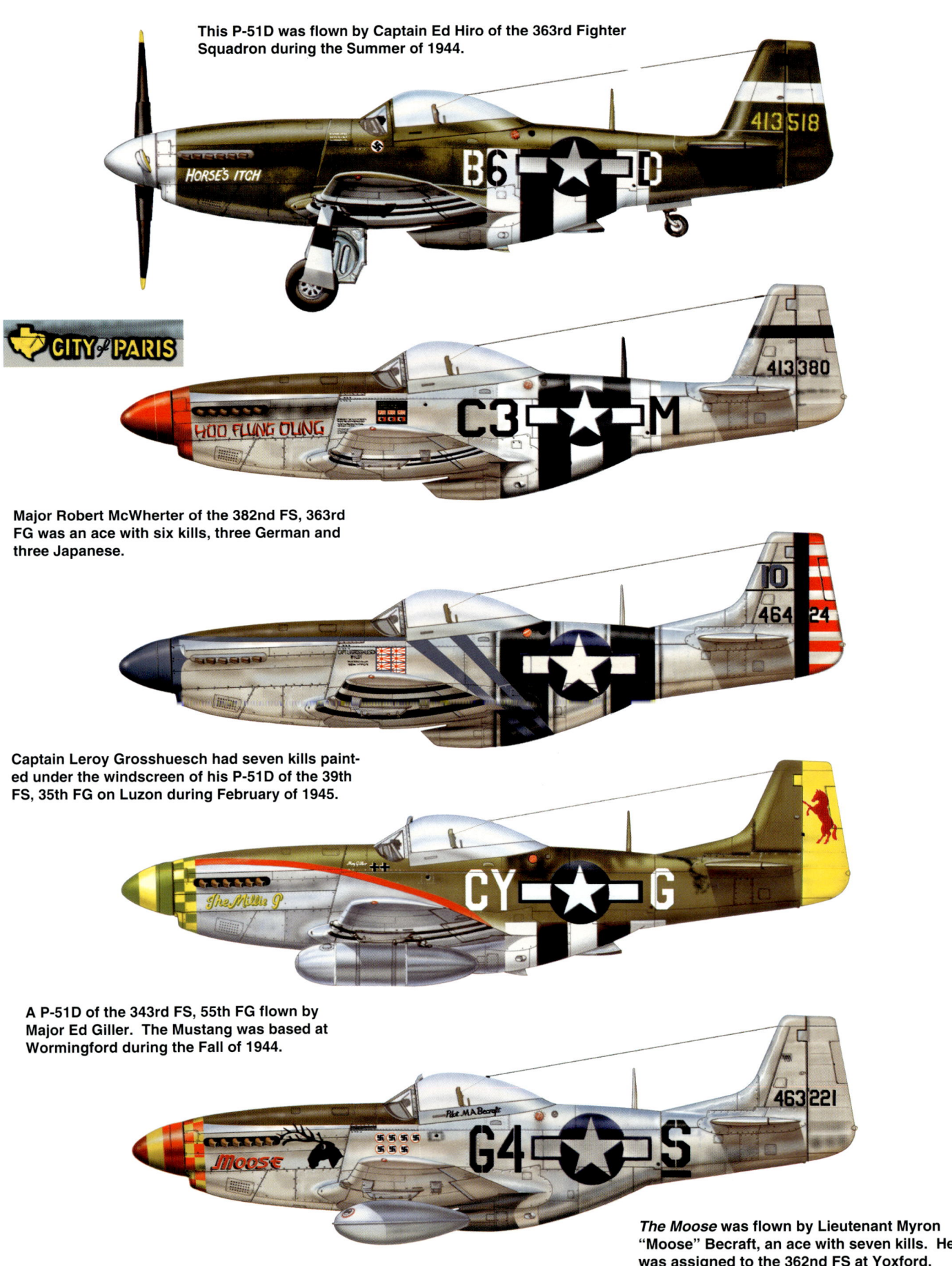

This P-51D was flown by Captain Ed Hiro of the 363rd Fighter Squadron during the Summer of 1944.

Major Robert McWherter of the 382nd FS, 363rd FG was an ace with six kills, three German and three Japanese.

Captain Leroy Grosshuesch had seven kills painted under the windscreen of his P-51D of the 39th FS, 35th FG on Luzon during February of 1945.

A P-51D of the 343rd FS, 55th FG flown by Major Ed Giller. The Mustang was based at Wormingford during the Fall of 1944.

The Moose was flown by Lieutenant Myron "Moose" Becraft, an ace with seven kills. He was assigned to the 362nd FS at Yoxford.

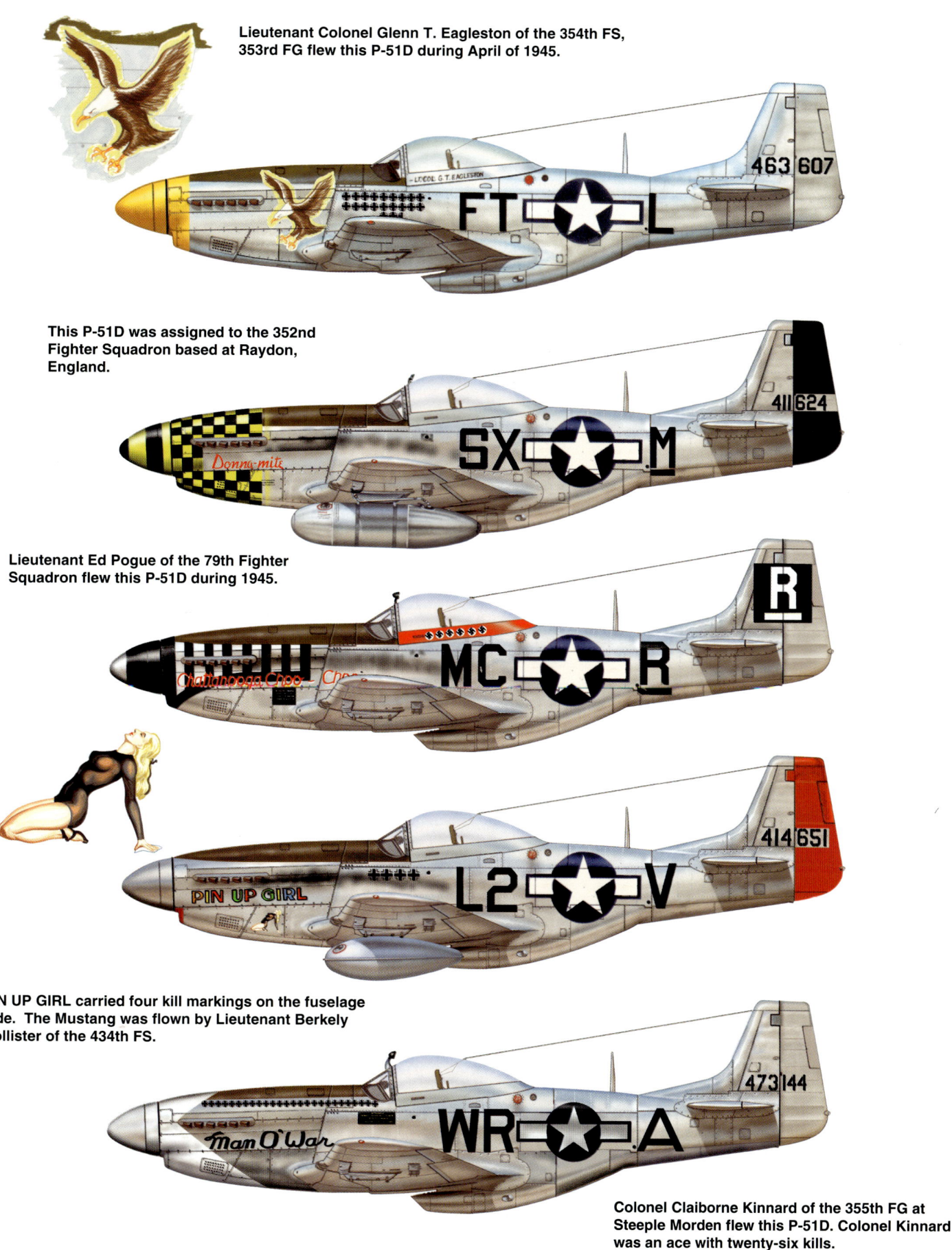

Lieutenant Colonel Glenn T. Eagleston of the 354th FS, 353rd FG flew this P-51D during April of 1945.

This P-51D was assigned to the 352nd Fighter Squadron based at Raydon, England.

Lieutenant Ed Pogue of the 79th Fighter Squadron flew this P-51D during 1945.

PIN UP GIRL carried four kill markings on the fuselage side. The Mustang was flown by Lieutenant Berkely Hollister of the 434th FS.

Colonel Claiborne Kinnard of the 355th FG at Steeple Morden flew this P-51D. Colonel Kinnard was an ace with twenty-six kills.

Bubbletop P-51D

The P-51D was the end result of all the improvements and developments in the basic P-51 design. It was powered by an improved 1,750 hp Packard/Merlin V-1650-7. Armament was improved by adding two more .50 caliber guns for a total of six in the wings. Additionally the guns themselves were mounted vertically in the gun bays instead of being angled as on previous models. This made the armament much more reliable and less prone to jamming.

But by far the biggest change was in the canopy design. North American engineers cut down the aft fuselage spine to a level even with the lower canopy sill; at the same time reinforcing the entire rear fuselage. A fully blown, sliding bubble canopy was fitted to the cut down rear fuselage. This full bubble canopy offered the pilot an unobstructed 360 degree vision - something all the Mustang pilots had craved for many years. Additionally the wing structure was also strengthened to be able to carry 1,000 pound bombs, or up to 165 gallon drop tanks.

P-51Ds began equipping the Mustang units beginning in the late Spring of 1944, with the 354th "Pioneer Mustang Group" again being the first unit to receive the new type. By the end of 1944, no less than forty-five squadrons in the U.S. 8th Air Force were equipped with Mustangs. Only the three squadrons in the 56th FG retained their P-47 Thunderbolts.

The Royal Air Force also began re-equipping with Mustang IV and IVA aircraft as soon as the new type was available. The RAF had 20 squadrons of Mustang IVs by the end of the war. Many of these units were equipped with the Dallas-built Mustang IVA (P-51K). The P-51K was equipped with an Aeroproducts propeller and had a slightly re-designed canopy shape, but otherwise it was identical to the P-51D. USAAF squadrons were also equipped with P-51Ks, often intermixed with P-51Ds in squadron use. Many of the squadrons; however, replaced the Aeroproducts propeller with the Hamilton Standard propeller due to manufacturing problems. The RAF took delivery of 282 Mustang IVs and 914 Mustang IVAs.

Reconnaissance versions of the P-51D/K had two or three cameras mounted in the rear fuselage and were designated F-6D and F-6K, depending on the base aircraft type used. The USAAF took delivery of 299 F-6D/K aircraft, equipping almost every reconnaissance unit during the war. Both Free French squadrons, Escadre 11-23 and Escadre 11-33, were equipped with F-6Ds or Ks.

By the end of the war North American Aviations two factories had built a total of 9,603 P-51D/K aircraft, equipping a full 75% of all fighter squadrons in the Army Air Force. They performed every mission profile called for by any fighter type aircraft. More Luftwaffe aircraft fell to Mustang guns than any other Allied type. This included a large number of the vaunted next generation fighter aircraft - the German jet and rocket planes. Major General Orvil Anderson, 8th Air Force Deputy Commander, summed it best, "It is my considered opinion that the P-51 played a decisive role in the air war over Europe!"

The ultimate in Mustang designs came about when North American Aviation modified the rear fuselage of this P-51B by removing the razorback and fitting the airplane with a full bubble canopy. This new canopy gave the pilot full 360 degree vision. (NAA)

Horse's Itch, flown by Captain Ed Hiro, was one of the first P-51Ds that were rushed to units in Europe. The aircraft was assigned to the 363rd Fighter Squadron at Yoxford during the Summer of 1944. (Merle Olmsted)

This P-51D from the 55th Fighter Squadron, 20th Fighter Group, 8th Air Force has had the upper surfaces camouflaged in a non-regulation pattern of Olive Drab.

Mom Smith's Little Angel served with the 364th Fighter Squadron at Yoxford. Although all P-51Ds were delivered in overall Natural Metal, the 357th Fighter Group camouflaged their aircraft in Olive Drab (or Medium Green 41) over Neutral Grey. (Merle Olmsted)

Fools Paradise IV was a P-51D of the 380th Fighter Squadron based at Azeville, one of the forward airstrips in France during August of 1944. The early P-51Ds were not fitted with the fillet at the base of the fin leading edge, that would become standard on later production P-51Ds. The upper wing D-Day invasion bands have been removed by "cleaning" with fuel-soaked rags. (USAAF)

Buzz Boy IV, was a 40th Fighter Squadron F-51D based at Irumagawa, Japan during 1948. The aircraft still carried its unit markings from the Second World War. (via Jeff Ethell)

A trio of sharkmouthed F-51D Mustangs of the *Grupo de Caza Ramfis* (Ramfis Fighter Group), *Fuerza Aérea Dominicana* (Dominican Air Force) on the ramp at Santo Domingo during 1966. (via Jeff Ethell)

F-51D Mustangs of the 109th Fighter Squadron/Minnesota Air Guard parked on the snow-covered ramp at Holman Field during 1954. (via Jeff Ethell)

The Spirit of Mactan was one of the Philippine Air Force Alert Flight aircraft at Nichols Field. These aircraft maintained a five minute alert against communist rebel forces during November of 1957. (Merle Olmsted)

40

DEE leads a mixed flight of 503rd Fighter Squadron Mustangs returning to Fowlmere from another mission to Big B (Berlin) during 1944. The near aircraft is an early P-51D without the fin fillet, the next two are late production P-51Ds with the fin fillet and the last aircraft is a P-51B fitted with a Malcolm Hood. (Don Garrett)

Captain Charles Weaver flew this unnamed, but nicely adorned P-51D when he was assigned to the 362nd Fighter Squadron at Yoxford. Captain Weaver was an ace credited with eleven victories. (Merle Olmsted)

HURRY HOME HONEY parked on the ramp at Leiston during July of 1944, was flown by Captain Richard Peterson. There are fourteen kill markings under the cockpit. (Keith Melville)

Gentleman Jim was flown by Captain James Browning. It carries the Red and Yellow spinner and checked nose common to aircraft of the 357th Fighter Group, 8th Air Force based at Yoxford, England. (Merle Olmsted)

Major General Claire Chennault talks with Sergant Walt Zarowski, a crew chief with the 51st Fighter Group at Kunming, China during 1944, about maintenance problems. ODI maintenance was almost always conducted on the ramp, regardless of weather conditions. (USAAF)

Major George Preddy, the top scoring Mustang Ace during the Second World War, waves from the cockpit of his Mustang, *Cripe's A' Mighty 3rd*. He was assigned to the 487th Fighter Squadron. The Mustang has eleven of his thirty-one kills painted on the nose. (USAFM)

(Right) Lieutenant Robert Black flew *Black's Bird*, while assigned to the 79th Fighter Squadron based at Kings Cliffe. Some 20th Fighter Group P-51s had Black and White stripes on the nose, while others had just Black stripes over Natural Metal. (USAFM)

Threebees, was a P-51D flown by Major John Burman with the 325th Fighter Group in Italy. The Yellow and Black checkerboard covers the entire vertical and horizontal tail surfaces and the top of the seventy-five gallon drop tanks. (via Dave McLaren)

(Above and Right) *Barbara* was flown by Captain Fred Haviland Jr. while assigned to the 357th Fighter Squadron. The markings on the Mustang changed as the war went on. During September of 1944, it had a White nose and rudder with ten kill markings. During November of 1944, it had a Blue nose and rudder with fourteen victory markings. (Bill Marshall)

Major Glenn Eagleston talks about the upcoming mission with other pilots from the 353rd Fighter Squadron in front of his P-51D. The Mustang has twenty and a half kill markings on fuselage behind the eagle. (Keith Melville)

Lieutenant Ed Winger flew "*Kiss Me*", a Yellow-nosed P-51D of the 41st Fighter Squadron based at Clark Field in The Philippines during the late Spring of 1945. (Fred Johnson)

Shack Lassie and another 383rd Fighter Squadron Mustang begin their takeoff roll from Honington during late 1944. The nose stripes of 364th Fighter Group P-51s were Blue and White. (via Jeff Ethell)

Molly shows four of Lieutenant "Cotton" Addis' victory markings painted under the canopy, and carries the full 325th Fighter Group Yellow and Black checkerboard on the tail and wingtips. (Warren Bodie)

Kathy, a P-51D from the 356th Fighter Squadron at Orconte, France, carries some nice artwork painted over the 'K' and through the 'Y'. (Colonel Richard Turner)

Tar Heel was flown by Captain James Starnes, an ace with six kills. He was assigned to the 505th Fighter Squadron based at Bassingbourne during 1945. (USAAF)

Captain Donald Bochkay touches down at Yoxford after another long range escort mission into Germany. Captain Bochkay had fifteen and a half victories while assigned to the 363rd Fighter Squadron. (Merle Olmsted)

A Black and White checkered nose P-51D from the 78th Fighter Group, sits on the ramp at Poltava following one of the Russia shuttle missions. These missions could last up to twelve hours, one way. (USAAF)

The *Chattanooga Choo Choo*, a P-51D assigned to the 79th Fighter Squadron at Kings Cliffe, carried six kill markings on the canopy rail. In the ETO, kill markings were added for both air and ground kills. The national insignia has been "grayed-out" and the bar under the code letter indicated it was the second squadron aircraft coded R. (Keith Melville)

Lieutenant Leo Elliot named his F-6D Mustang *Tampa Joe*. He was assigned to the 12th Tactical Reconnaissance Squadron of the 10th Photo Reconnaissance Group at St. Dizier, France during 1944. (Leo Elliot)

Lieutenant Colonel Ernest Beverly flew this Yellow and Black checkered Mustang when he commanded the 325th Fighter Group at Lesina, Italy during 1944. Lieutenant Colonel Beverly scored three victories. (Ernie McDowell)

Lieutenant Colonel John Meyer's famous *Petie 2nd*, parked on the ramp at Bodney during 1945 carried twenty kill markings under the cockpit. Lieutenant Colonel Meyer ended the war with twenty-four victories. (via Tom Ivie)

The last of Captain Clarence Anderson's *Old Crow*'s sits on the snow-covered 363rd Fighter Squadron ramp at Yoxford with eighteen victory markings on the fuselage under the windscreen. The aircraft also carried two 110 gallon pressed paper drop tanks. (Merle Olmsted)

Patty IV, was a P-51D flown by Lieutenant Lawrence McCarthy while assigned to the 328th Fighter Squadron. The aircraft is equipped with a pair of 75 gallon metal drop tanks under the wings. Lieutenant McCarthy scored two and a half victories. (Keith Melville)

Major Robert McWherter, Commanding Officer of the 363rd Fighter Squadron, claimed three Japanese victories while with the 17th Pursuit Squadron, and three German victories while with the 363rd Fighter Squadron. His Mustangs were all named *City Of Paris/Hoo Flung Dung*, and had a map showing the location of Paris, Texas. (Don Garrett)

Pauline was flown by Lieutenant Colonel Joe Thury when he was assigned to the 505th Fighter Squadron at Fowlmere during February of 1945. His Mustang had two Spitfire-type rear view mirrors and carried 110 gallon drop tanks. (USAAF)

Madam Wham Dam, a 458th Fighter Squadron P-51D, parked in its revetment at North Field on Iwo Jima during 1945. The gun ports were taped over to retard dirt and corrosion in the gun barrels. (Ron Witt)

The *Big Noise From Winnetka* was flown by Flight Officer Thomas Bur of the 369th Fighter Squadron based at East Wretham during August of 1944. (A.C. Chardella)

A P-51D from the 45th Fighter Squadron taxiis back to the parking ramp on South Field on Iwo Jima. Mustangs based on Iwo Jima could easily range over the entire Japanese Home Island chain and escorted B-29s in their raids on the Japanese homeland. (USAAF)

F-6D Mustangs from the 110th Tactical Reconnaissance Squadron line the ramp on Ie Shima during August of 1945. The aircraft have canvas covers over the cockpits to hold back some of the oppressive tropical heat. The Black and White bands on the wings and fuselage were a 110th Tactical Reconnaissance Squadron marking. (via Joe Bruch)

(Above) Certainly the flashiest paint schemes seen on combat Mustangs were those of the 343rd Fighter Squadron, which had the uppersurfaces of their Mustangs camouflaged in a stylized Olive Drab. Captain Robert Welch scored six victories in *Miss Marilyn II*. (Keith Melville)

Tiny "Gay Babe" and *Three Of A Kind*, a pair of 46th Fighter Squadron Mustangs, parked on the North Field ramp during 1945. *Three Of A Kind* was named because its serial number ended with "777". (USAAF)

Captain Richard Peterson returns to Leiston in his last Mustang named *HURRY HOME HONEY*, during the Fall of 1944. Captain Peterson scored a total of nineteen victories. (Keith Melville)

Penny 4 was the Mustang assigned to Colonel John Lowell, a seventeen victory ace with the 384th Fighter Squadron at Honington during 1945. (USAAF)

A batch of new 69th Photo Reconnaissance Group F-6D Mustangs get the final check by technicians at Speke during the Summer of 1944. The Mustangs left Speke carrying complete squadron markings, colors, and code letters. (USAAF)

The 7th Air Force had several groups of long range Mustangs based on the island of Iwo Jima. Their mission was escort of the B-29s to targets on the Japanese Home Islands. *Pee Wee* and these other 78th Fighter Squadron P-51Ds and P-51Ks were assigned to the 15th Fighter Group. (USAAF)

P-51D Mustangs from the 506th Fighter Group line up for takeoff from North Field on Iwo Jima during the Summer of 1945. Operations on Iwo Jima were always accompanied by a cloud of volcanic dust. (USAAF)

DooleyBird, was a Mustang IV flown by Flight Lieutenant A.S. Doley of No 19 Squadron during 1945. The Mustang IV was the RAF designation for the P-51K. (A.S. Doley via R.L. Ward)

Jill's Jalopy III carries the early 1945 markings of the 343rd Fighter Squadron, which consisted of a simple Yellow and Green propeller spinner and checkered nose, with a Yellow rudder.

Sigh!, a P-51K assigned to the 25th Fighter Squadron at Kunming during 1945, carries a Black checkerboard on the tail, the unit marking of the 25th Fighter Squadron. The five Black stripes on the fuselage identify the aircraft as having been assigned to the 1st ACG at one time. (Keith Melville)

(Left) A P-51K from the 383rd Fighter Squadron returns to Honington during 1944. The P-51K was a Dallas-built P-51D that had a reshaped canopy and AeroProducts propeller. (J. Fields)

Donna-mite, a P-51D of the 352nd Fighter Squadron, lands at Raydon, England, following another bomber escort mission into Germany. The lack of underwing drop tanks on the Black and Yellow checkered Mustang, indicates that the Luftwaffe was encountered. (USAAF)

Members of the 5th Fighter Squadron spelled out the word "VICTORY" using the squadron code letters painted on the tails of their P-51Ds at Madna, Italy during 1945. (David Weatherill)

Captain James Duffy scored a total of fourteen air and ground victories in *DRAGON WAGON,* a 354th Fighter Group P-51D. The 8th Air Force was the only Army Air Force to give credit for ground victories. (355th FG Assn.)

Major Jack Ilfrey and his ground crew stand near *Happy Jack's Go Buggy*, the P-51D he flew while attached to the 79th Fighter Squadron. His personal scoreboard was painted on the nose and included six kills. (USAAF)

This P-51K from No 3 Squadron, Royal Australian Air Force, rests on a PSP parking ramp in Italy during 1945. The P-51K has a Red spinner, code letters, and rudder, with White stars. British Mustang IV units were almost exclusively equipped with the P-51K. (Don Garrett)

Major Bill Dunham was credited with sixteen Japanese aircraft, two Japanese ships, and thirty bombing mission. He flew a P-51K while he commanded the 460th Fighter Squadron on Luzon during 1945. (Dwayne Tabatt)

MAN O' WAR, Lieutenant Colonel Claiborne Kinnard's P-51D had twenty-six kill markings painted on the nose. He commanded the 355th Fighter Group during February of 1945. (Robert Kuhnert)

The Philly Pip, a P-51D from the 4th Air Commando Squadron, flies off the coast of Okinawa during 1945. Every island that was captured meant another base closer to the Japanese Home Islands. (Paul McDaniel)

Captain Ray Littge smiles from the cockpit of his 487th Fighter Squadron P-51D. The aircraft has twelve of his total of twenty-three and a half kills painted on the canopy frame. (USAAF)

DADDY'S GIRL was the personal mount of Captain Ray Wetmore of the 359th Fighter Group, an ace with twenty-one kills. (Keith Melville)

Ground crew personnel push a Yellow-tailed P-51D from the 462nd Fighter Squaron to its revetment at North Field on Iwo Jima during 1945. The antenna on the vertical fin was a tail warning radar. (USAAF)

Contrary Mary was flown by the Commanding Officer of the 78th Fighter Group at the end of the war, Lieutenant Colonel Roy Caviness. British civilians crowd around the beautiful Black and White checkered Mustang on Air Force Day at Duxford, 1 August 1945. (USAAF)

Bulldogs/Jane III was flown by Captain Bert Marshall when he commanded the 354th Fighter Squadron, which was denoted on his aircraft by the letter "C" on the rudder. (Don Garrett)

Lieutenant Myron "Moose" Becraft flew this P-51D named *Moose*, He had seven victories in the 362nd Fighter Squadron at Yoxford during 1945. (Merle Olmsted)

American Maid, Lieutenant Eugene Sears' P-51D, sitting at its parking spot at Wattisham, England, home plate for the 434th Fighter Squadron during 1945. (Keith Melville)

Corky was a 79th Fighter Squadron P-51D. The aircraft was parked on the ramp at Kings Cliffe, England. The 20th Fighter Group had always been a long range escort unit and had flown Lockheed P-38 Lightnings before converting to the P-51 during 1944. (Don Garrett)

The Millie G, with Major Ed Giller at the controls, reveals the fancy camouflage and markings common to Mustangs of the 343rd Fighter Squadron during the Fall of 1944. The fuselage was painted Olive Drab with Red trim, a Green and Yellow spinner and checkerboard on the nose, with a Yellow rudder. (Keith Melville)

Major Robert Moore walks to his 15th Fighter Group P-51D named *Stinger VII*. Major Moore had twelve confirmed Japanese victories while flying from Iwo Jima during 1945. (USAAF)

Lieutenant Evan Johnson was a five victory ace. He flew *THE COMET*, a P-51D of the 505th Fighter Squadron during June of 1944. (USAAF)

Lieutenant Louis Curdes took his toll of Axis aircraft during the war. Lieutenant Curdes had one Italian victory, one Japanese, seven German, and a U.S. C-47 that he shot up to prevent its capture by the Japanese in April of 1945. (USAAF)

Frank Bertciel's *Miss Velma* carried the stylized camouflage applied to 343rd Fighter Squadron P-51Ds, which even covered the serial number. The underlined code letter indicated it was the second squadron aircraft with the letter D. (Keith Melville)

William Odom's highly modified overall Dark Green P-51B Beguine, had the radiators moved to the wingtips. The aircraft was flown in the 1949 Thompson Trophy race and crashed, killing Odom, a women and her thirteen year old son. (NAA)

Linton Carney flew the Houstonian at the 1948 Cleveland National Air Races, averaging over 446 mph, for a second place finish. The aircraft was owned by Paul Mantz. This was the same aircraft in which Mantz won the 1946 Bendix race. (NAA)

Many racing Mustangs are painted in a semi-realistic Second World War paint schemes, like Gerald Martin's race number 12. It is painted with Red and Blue diamonds, like those found on 356th Fighter Group P-51s. (Don Garrett)

(Left) Paul Mantz's other P-51C was the Red and White #46, in which he won the 1946, 1947, and 1948 Bendix Trophy races, averaging as high as 460 mph. The aircraft had a sealed "wet" wing giving it additional fuel for the long race. (NAA)

Other racing Mustangs carry markings that bear no resemblance to any real unit markings. This P-51D is painted in Vietnam War style camouflage, has early Second World War national markings, and carries shark's teeth from the China-Burma-India Theater. (Don Garrett)

Danny Boy is another privately owned Mustang with a conglomerate of markings. It has the correct style of camouflage, accurate D Day bands, and the Yellow nose of an aircraft from the 361st Fighter Group, but the fuselage codes are incorrect. (NAA)

A camouflaged P-51D of the Israeli Defense Force/Air Force. Mustangs flew in several of the Arab/Israeli wars, being used mainly in the ground support role. (via Hans-Heiri Stapfer)

Personnel from No 2 Squadron/SAAF build a sandbag revetment around one of their F-51D fighter-bombers. The squadron was attached to the 18th Fighter Bomber Group during 1952. (USAF)

One of the many countries that re-equipped with Mustangs after the end of the Second World War was Sweden. Designated the J26 in Swedish service, this Mustang had been assigned to an Israeli squadron before being sold to Sweden. (Jim Sullivan)

This F-51D was assigned to the 1st Fighter Squadron, Philippine Air Force, and was based at Nichols Field during December of 1957. The Mustang is equipped with Fletcher drop tanks originally designed for the F-80/T-33 Shooting Star. (Merle Olmsted)

Most of the Central American and Caribbean nations were equipped with Mustangs as their primary fighter aircraft during the 1960s. This Dominican Republic F-51D was camouflaged in a three tone camouflage of Medium Green, Dark Green and Tan (USAF Vietnam War style) and had miniature national insignia. (via Nick Waters)

North Korean anti-aircraft fire could be quite accurate, especially against the low-flying reconnaissance aircraft like this RF-51D from the 15th Tactical Reconnaissance Squadron. The aircraft had twenty-six mission markings on the fuselage below the windscreen. (Major General Stanley Newman)

The Mustang was more than a match for the North Korean Yak fighters. Captain Alma Flake (left) and Lieutenant James Glessner (second from right) each shot down a YAK-9 on 2 November 1950, flying with the 12th Fighter Bomber Squadron. (USAF)

Colonel Dean Hess flew this F-51D when he commanded the BOUT ONE force at Taegu, South Korea. The legend on the nose translates to read " By Faith I Fly." (Keith Melville)

No 77 Squadron/RAAF pilots are greeted by officers from the U.S. 5th Air Force after their arrival in Korea during November of 1950. The Royal Australian Air Force squadron was initially assigned to the 6150th Tactical Support Wing. (USAF)

The winter in Korea was the most brutal any of these South African crew chiefs had ever seen. Based as far north as Pyongyang, North Korea, No 2 Squadron/SAAF was forced to retreat back down the peninsula before the onrushing Chinese hordes. (USAF)

Under Project BOUT ONE, the U.S. supplied South Korea with one squadron of veteran F-51D Mustangs. Under the command of Colonel Dean Hess, USAF, the Republic of Korea Air Force (ROKAF) squadron was trained by U.S. Air Force pilots, who also flew many combat missions in the ROKAF Mustangs. (U.S. Marine Corps)

The F-51D was the primary fighter-bomber aircraft in Korea. These trailers full of 5 inch HVAR rockets are waiting to be loaded onto Mustangs at Chinhae during November of 1951. Mustangs from three of the four squadrons assigned to the 18th Fighter Bomber Wing are visible, No 2 Squadron/SAAF, 12th Fighter Bomber Squadron and 39th Fighter Bomber Squadron. (USAF)

Helen San, a 40th Fighter Interceptor Squadron F-51D was part of the Japan Air Defense Force when the 35th Fighter Interceptor Group's F-80C Shooting Stars were needed in Korea. (Dave McLaren)

Doris + 2/Swing Lo carries the smiling sharkmouth painted on aircraft from the 12th Fighter Bomber Squadron at Chinhae during 1952. (Jeff Ethell)

The other United Nations unit flying Mustangs during the Korean War was No 77 Squadron, RAAF. They flew fighter-bomber missions from Taegu during August 1950. No 77 Squadron transitioned to Meteor jets in July of 1951. (U.S. Army)

When the Korean War broke out in June of 1950, the only fighters available in quantity for combat were F-51Ds in storage in the Far East. The 12th Fighter Bomber Squadron and the fledgling Republic Of Korea Air Force (ROKAF) flew F-51Ds from K-2, Taegu, during August of 1950. (U.S. Army)

Reconnaissance of the North Korean road network was the mission of the RF-51Ds from the 67th Tactical Reconnaissance Squadron. *Sooner Snooper* carried the Blue and White polka dot markings of the 45th Tactical Reconnaissance Squadron. (Major General Stanley Newman)

Ol' NaDSOB, a 67th Fighter Bomber Squadron F-51D taxis through one of the many "lakes" at Chinhae during September of 1951, The Mustang was loaded with two 500 pound bombs and four 5 inch HVAR rockets. (USAF)

Maintenance in Korea was always under the crudest of conditions. These 12th Fighter Bomber Squadron mechanics use a portable air compressor during the 100 hour maintenance check of this F-51D at Chinhae during 1952. (USAF)

Several changes were made to these 165th Fighter Bomber Squadron/Kentucky ANG F-51Ds, including a fixed, non-retractable tail wheel, which was retrofitted to many Mustangs during the early 1950s. (ANG/HO)

A TF-51D two-seat trainer assigned to the 113th Fighter Bomber Squadron/Indiana ANG during 1953. The second seat in a TF-51D took the place of the fuselage fuel tank. (Dave McLaren)

These 195th Fighter Squadron/California ANG F-51Ds were painted to portray "Messerschmitt 109s" for the 1948 Hollywood war film "Fighter Squadron," which starred P-47 Thunderbolts. (CA ANG)

A very colorful F-51D assigned to the Fighter School at Luke Air Force Base during the early 1950s. The markings are Gold and Black. (NGB)

California Air Guard F-51Ds of the 195th Fighter Squadron on the ramp at Van Nuys Airport during 1950. Some are originally P-51Ds, others are P-51Ks. But all have the all-metal rudder, ailerons, and elevators retrofitted after the war. FF-631 has just been transferred in from an active Air Force unit and carries no Guard markings. (NAA)

Mechanics prepare to warm up *Betty Jo*'s big Merlin at Chitose. *Betty Jo* was assigned to the 36th Fighter Squadron during the Winter of 1948. (USAF)

Many Air National Guard squadrons were activated during the Korean War. These 163rd Fighter Squadron, Indiana ANG pilots scramble during an alert at Fort Wayne. (Dave McLaren)

Many stateside Air Defense Command units, like the 37th Fighter Interceptor Squadron, were still equipped with Mustangs well into the 1950s, although the "threat" would have been from a jet-powered enemy. (Marty Isham)

This P-51 of the Swiss Air Force carried the early style of national insignia which consisted of a Red rudder with a White cross, and Red squares on the wings with White crosses. The only other markings carried was the serial number which was in Black on both sides of the fuselage. (Hans-Heiri Stapfer)

As F-51Ds were phased out of front line Air Force squadrons, they were phased into service with the Air National Guard. This flight of F-51Ds was assigned to the 116th Fighter Interceptor Squadron during 1953. (Washington ANG)

Mustangs were also the main fighter aircraft of many Allied air forces well into the 1950s, including the Royal Australian Air Force. Some of the RAAF Mustang IVs were license-built by Commonwealth Aircraft Corporation in Australia. (USAF)

(Right) A line-up of F-51Ds of 6 Stormo, Italian Air Force during 1950. Later, a number of Italian F-51s were passed to the Somalian Air Force. (via Hans-Heiri Stapfer)

This "pranged" F-51D of the Swiss Air Force carries the late style national insignia. (Hans-Heiri Stapfer)

Dee was a F-51D from the 39th Fighter Squadron based at Irumagawa during 1948. The fancy scallop markings on the nose and wing tips were Dark Green. (Keith Melville)

F-51Ds from the 27th Fighter Squadron pass by Mount Rainier, Washington during 1947. The band on the rear fuselage is Medium Blue and the spinners were Red. (Don Maggert)

Bowlin' Baby was flown by Major J.P. Carson, and still carries the Second World War 40th Fighter Squadron markings. The aircraft was based at Irumagawa during 1948. (Keith Melville)

The Post-War Years

When the Second World War ended, many of the foremost combat fighter aircraft types were unceremoniously carted off to various aircraft junk yards and scrapped! P-38s, P-40s, and P-47s of all types, some being flown directly from the assembly lines, were parked and destroyed or sold for scrap. Not so with the P-51s. Except for the War-Weary P-51B/C types, the Mustang would be the backbone of the post-war United Nations air forces. Most of the USAAF, and later USAF, squadrons were equipped with P-51Ds, even though this was the dawn of the jet age. The RAF, RCAF, Italian, French, Chinese, and many other Allies all were equipped with Mustangs as their primary air defense weapon in the late 1940s.

Inevitably, as more and better jet aircraft entered the inventories, the P-51s were relegated to a reserve status. The first Mustangs to enter service with the Air National Guard were those assigned to the 120th FS/Colorado ANG, going operational during June 1946. By the end of Mustang operations in U.S. service during 1957, some seventy Air National Guard squadrons had been equipped with the (now) F-51D.

Active U.S. Air Force units were still flying the F-51D well into the 1950s. Part of this was due to the call-up of many jet units to the "Police Action" in Korea. When the North Korean tanks crossed the border in June of 1950, the U.S. Air Force needed as many aircraft as they could get - and they needed them RIGHT NOW! Mustangs were plentiful, especially in the Far East where entire squadrons had simply parked their aircraft following conversion to jets. Some units, like the 8th FBG at Itazuke, found themselves converting back to F-51Ds from Lockheed F-80 Shooting Stars. The reason was because they needed an aircraft with more "loiter time" over the target than the F-80 jets could provide. The answer was the F-51D Mustang.

In Korea the U.S. 5th Air Force had three groups of F-51D fighter-bombers, and one squadron of RF-51D reconnaissance aircraft, available at a moments notice. Two additional UN squadrons had one squadron each of Mustangs, No 77 Squadron, RAAF and No 2 Squadron, SAAF. Plus the fledgling Korean Air Force was equipped with F-51Ds, although they were being flown by USAF pilots. Back in the U.S. as the war heated up, more and more jet aircraft were needed in Korea to fight the Soviet MiGs. These jets were replaced in their stateside units by F-51Ds. Additionally, many of the "activated" Air Guard squadrons began flying primary Air Force missions, such as air defense duty, with F-51Ds. It wasn't until March of 1957 that the last F-51 mission was flown.

But that was hardly the final Mustang mission. During 1956 Israel found itself embroiled in another war. The primary aircraft in the Israeli Hey'l Ha'Avir was the F-51D fighter bomber. In 1959, Philippine Air Force F-51Ds fought against communist rebel forces in the hills of northern Luzon. Most of the Central American nations had air forces equipped with F-51Ds that were engaged in a seemingly endless string of border wars. The Mustang was born to fight and only technology seems to have been able to erode its capabilities to do just that. But who knows what the future will bring. Only recently there was still another proposal to bring back the so-called SUPER Mustang, an aircraft of the past fighting the brushfire wars of the future.

Mustangs were the backbone of the occupational air forces in both Germany and Japan. *Vivian* **was assigned to the 80th Fighter Squadron based at Chitose, Japan during the Winter of 1948. (USAF)**

Air Force stateside squadrons were still equipped with Mustangs well into the 1950s. Now designated as a F-51D, this Mustang took part in the Army/Air Force Maneuvers held at Wilmington Airport, North Carolina during 1951. (Paul McDaniel)

Miss Steve rests on her belly after a crash landing. The aircraft carried the kill tally (eighteen) of Lieutenant Bill Cullerton, one of the many aces in the 357th Fighter Squadron. (355th FG Assn.)

This P-51D from the 47th Fighter Squadron was unloaded from the USS SITKOH BAY during February of 1945, then transported by motor barge to the beach on Iwo Jima. (USAAF)

A Mustang IVa (P-51K) assigned to No 3 Squadron, Royal Australian Air Force during 1945. The rudder was Red with White stars. (via Jeff Ethell)

Lieutenant Colonel Donald Baccus flew this Green-nosed P-51D when he commanded the 359th Fighter Group during 1945. His scoreboard shows eight victories. (A.C. Chardella)

Typical end of the war markings were carried on this 383rd Fighter Squadron Mustang. *BABS IN ARMS* was based at 8th Air Force Station F-375 (Honnington), England during May of 1945. (USAAF)

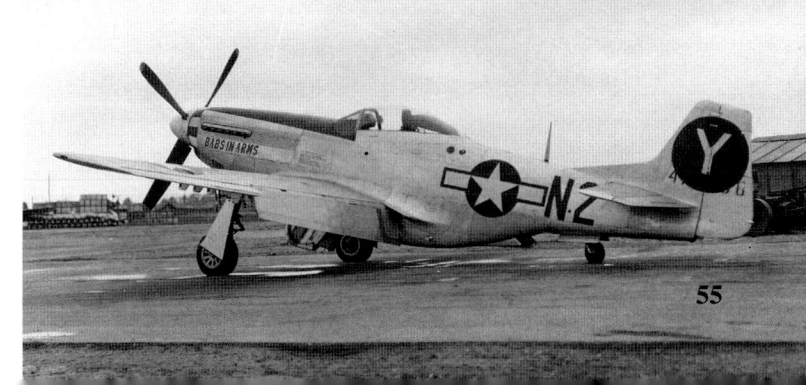